THE ICE AGE
AND THE FLOOD

Does Science Really Show Millions of Years?

JAKE HEBERT

INSTITUTE FOR CREATION RESEARCH

Dallas, Texas
www.icr.org

Dr. Hebert earned a bachelor's degree in physics in 1995 from Lamar University and a master's degree in physics in 1999 from Texas A&M University, where he studied optics and was a Dean's Graduate Fellow 1995-1996. He received his Ph.D. in 2011 from the University of Texas at Dallas, where his research involved a study of the possible connection between fair-weather atmospheric electricity and weather and climate. He joined ICR in 2011 as a research associate, where he has had the opportunity to help extend Dr. Larry Vardiman's work on climates before and after Noah's Flood, among other research endeavors.

THE ICE AGE AND THE FLOOD
Does Science Really Show Millions of Years?

by Jake Hebert, Ph.D.

First printing: July 2014

ISBN: 978-1-935587-65-1

Please visit our website for other books and resources: www.icr.org

Printed in the United States of America.

TABLE OF CONTENTS

INTRODUCTION

Most people are familiar with the concept of an ice age, a time when large ice sheets and glaciers covered much of the earth's land surface. There is strong geological evidence that today's high-latitude ice sheets once extended to much lower latitudes. In North America, glaciers once covered nearly all of Canada and even extended as far south as Kansas. Mysterious creatures, such as Neandertals and the now-extinct wooly mammoths, lived at this time. What could cause an ice age? Was there more than one? And how does this fit into biblical history?

Secular scientists believe there were at least five major ice ages in Earth history. In fact, some claim that nearly the entire earth was frozen about 650 million years ago. The most recent of these major ice ages is said to have begun at the start of the Pleistocene Epoch about 2.6 million years ago. If one defines an ice age as a period during which large ice sheets are present on Earth, then according to that definition we are still in an ice age. However, what most people think of as an ice age is actually called a glacial interval (or *glacial*) by the specialists. A glacial is a period during which the ice sheets advance and become considerably larger.

Based on their interpretations of chemical data from cores drilled and extracted from deep seafloor sediments and high-latitude ice sheets, secular scientists think that more than 50 of these glacial ice ages (of varying sizes) have occurred within the last 2.6 million years.[1] These glacials are said to have been separated from one another by relatively warmer periods called *interglacials*. Within the last 750,000 years, about nine glacial-interglacial cycles have supposedly occurred.

As you can see, secular discussions of ice ages are closely connected to the idea of "millions of years" and a very old earth. This belief in deep time is based on claims such as:

- Hundreds of thousands of yearly layers have supposedly been identified within ice cores extracted from the Greenland and Antarctic ice sheets. For instance, ice from one particular Greenland ice core (the GISP2 core) at a depth of 2,800 meters is said to have been deposited about 110,000 years ago.[2]

- Seafloor sediments were slowly deposited over many millions of years.

- Chemical clues from cores drilled in both the ice and seafloor sediments record information about climate in Earth's "prehistoric" past and tell consistent stories about how Earth's climate has varied over hundreds of thousands, and even millions, of years.

However, a straightforward reading of the Bible indicates a young age for the earth—about 6,000 years.[3] Because of the apparent conflict between secular claims and the Bible's account of Earth history, many Christians have felt intense pressure to re-interpret Genesis to allow for these supposed millions of years.

How should Christians respond to these claims? Was there really an ice age or ages? If so, how many? Are the Greenland and Antarctic ice sheets really millions of years old? Have seafloor sediments been deposited slowly over millions of years? And if the ice cores and seafloor sediments are *both* "telling" a consistent story about how climate has varied for hundreds of thousands of years, isn't this an extremely strong argument for an old earth?

This booklet summarizes the key points regarding creation explanations for the Ice Age, particularly as they relate to the timing of the age and the dating of the deep ice cores. It draws on the work of other creation scientists, particularly Larry Vardiman and Michael Oard, to show that the Bible holds the key to understanding Earth's past.[4]

1

SEAFLOOR SEDIMENTS, ICE CORE COUNTING, AND THE ASTRONOMICAL THEORY

Before we can really begin discussing the Ice Age, we must first lay some groundwork. To see why the secular scientists' ice-age theories don't work, we need to understand the approach they use in their dating methods. This can get a bit technical, but it allows you to see the flaws in their methods and models—especially compared to the wonderfully straightforward model based on the Genesis account.

Two Approaches to Earth History

When we consider Earth's history, it is extremely important to remember that, strictly speaking, science deals only with the observable present. Scientists cannot travel back in time to observe the past. However, like crime scene investigators, scientists can examine clues and attempt to reconstruct historical events. Unlike the *observational sciences* such as physics and chemistry, conclusions drawn by researchers in the *historical* or *forensic sciences* are much more likely to be influenced by the researchers' starting assumptions.

Creation and secular scientists make dramatically different assumptions about the past. Secular scientists start by assuming that the Bible's account of Earth history is not true, and they interpret the data according to this premise. For instance, we find fossil remains of billions of organisms buried in water-deposited rock layers all over the earth. These creatures had to have been buried rapidly or their bodies would have decayed or been eaten by scavengers before they could become fossilized. The remains of billions of organisms in water-deposited rock layers is exactly what one would expect if the Genesis Flood occurred as described in the Bible (Genesis 6–9). Yet secular scientists insist from the outset that such an interpretation of the data is invalid.

Secular scientists hold to a philosophy called *uniformitarianism*, which is summarized by the motto "the present is the key to the past." This philosophy denies God's intervention in history. It also claims that mutations and natural selection can explain the existence of life on Earth apart from a supernatural creator. Likewise, it claims that geologic processes, operating at generally the same rates and intensities as they do now, can explain the rocks and fossils apart from the global Flood described in the Bible. Uniformitarians deny God's creation of the world and God's judgment by the Flood of the world in the days of Noah, and they deny that Jesus Christ will one day return to judge the world and set up His kingdom. The apostle Peter prophesied that in the last days "scoffers" would come who would hold to such a philosophy:

> . . . knowing this first: that scoffers will come in the last days, walking according to their own lusts, and saying, "Where is the promise of His coming? For since the fathers fell asleep, all things continue as they were from the beginning of creation." For this they willfully forget: that by the word of God the heavens were of old, and the earth standing out of water and in the water, by which the world that then existed perished, being flooded with water. (2 Peter 3:3-6)

Creation scientists, on the other hand, start from the premise that the Bible is God's inerrant Word and use that to interpret the scientific and historical data.

There are two important points to remember about the Ice Age:

1. Secular (uniformitarian) theories cannot explain an ice age.

2. The Bible, on the other hand, provides an extremely convincing explanation.

We will return to these important points later. However, in order to answer secularists' claims about multiple ice ages over millions of years, it is necessary to take a detour. We must first discuss the way that uniformitarian scientists interpret seafloor sediments and ice cores, as well as something called the *astronomical theory* of ice ages.

Slow Deposition of Seafloor Sediments

At first, seafloor sediments might seem to have absolutely no connection to an ice age, but they are necessary for understanding the long ages assigned to ice cores. At today's slow rates, a thousand years of deposition are needed to deposit a few centimeters of sediment on the deep ocean floor. Scientists have drilled and extracted cylindrical cores that can have total lengths of many hundreds of meters. Given the slow deposition rates, uniformitarian scientists assume that millions of years were required for these sedimentary layers to have formed.

Of course, one would expect sedimentation rates to be much higher during and shortly after the Genesis Flood.[1] However, secular scientists who hold to a uniformitarian philosophy do not consider this since they reject the possibility of a global worldwide flood. Instead, they believe that the chemistry of the seafloor sediments can provide information about climates millions of years ago. For example, they think that the amount of one kind of oxygen atom compared to another kind of oxygen atom can tell them about the advance and retreat of ice sheets in the "prehistoric" past.

Oxygen Variations within Seafloor Sediment Cores

To understand the significance placed on the seafloor sediments, we must delve into the realm of chemistry. There are two fairly common varieties, or *isotopes*, of the oxygen atom. One of these, oxygen-16, is about 500 times more common than the heavier variety of oxygen called oxygen-18. A quantity called the *oxygen isotope ratio* measures the amount of oxygen-18 compared to oxygen-16. Higher values of this ratio indicate an increased amount of oxygen-18 relative to oxygen-16 (compared to a standard), while a smaller value implies decreased amounts of oxygen-18. In scientific papers, this oxygen isotope ratio is often abbreviated as the symbol $\delta^{18}O$.

During their lifetimes, tiny marine organisms called Foraminifera (forams for short) build shells made of calcium carbonate ($CaCO_3$), which includes oxygen. Forams use *both* varieties of oxygen to construct their shells. When these organisms die, their shells drift downward and become part of the ocean sediments. From the remains of these shells, researchers can determine the oxygen isotope ratio at different depths within the cores that have been extracted from the seafloor sediments. Uniformitarian scientists believe that the oxygen isotope ratio gives them information about how climate varied in the distant past. When these values are plotted on a graph, one sees many "wiggles" as the oxygen isotope ratio goes up and down. High values of the oxygen isotope ratio within the sediments are thought to indicate ice ages (Figure 1). It should be noted that it is also possible to calculate $\delta^{18}O$ values for seawater, since water contains oxygen.

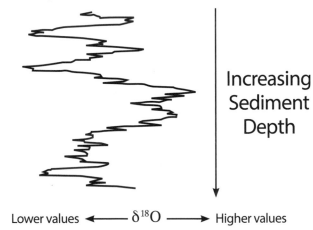

Increasing Sediment Depth

Lower values ⟵ $\delta^{18}O$ ⟶ Higher values

Figure 1. *Secular scientists believe that "wiggles" in the chemistry of the seafloor sediments can yield information about past climates. For example, maximum values in a quantity called the oxygen isotope ratio ($\delta^{18}O$) are thought to indicate times of maximum glacial extent.*

However, there are serious difficulties in attempting to infer past climates from foram $\delta^{18}O$ values. One particularly serious problem is that the foram $\delta^{18}O$ value depends upon both the temperature and $\delta^{18}O$ value of the seawater at the time the foram shell was formed.[2] The present oxygen isotope ratio of the foram shell can be measured in the laboratory, but there is no way for a scientist to travel back in time to measure either of these other two values. This means that uniformitarian scientists have to make assumptions in order to fill in this missing information. For these and many other reasons, inferring information about past climates from the chemistry of seafloor sediments is *extremely* problematic.[3]

Since secular scientists claim that seafloor sediments were deposited over millions of years, you may wonder how uniformitarian scientists date a layer of seafloor sediments. Radioisotope methods generally can only be used to date volcanic rocks, not sediments, so secular scientists must use other methods to attempt to assign ages to the sediments. The graph in Figure 1 shows how the oxygen isotope ratio varies with sediment depth, but *not* with time. So how do secular scientists convert depths of seafloor sediments into ages?

Keep this question in mind as you continue to read; it becomes important later in our discussion. If this section seemed a little technical, don't worry. The main point is that secular scientists think that they can accurately identify the layers of sediments that were deposited during ice ages by examining chemical "clues" within those sediments.

The Astronomical (Milankovitch) Theory of Ice Ages

To consider the most popular theory of ice ages, we must move from chemistry to astronomy. The earth rotates once every 24 hours around an imaginary line called its *axis of rotation*. Right now, the earth's axis is tilted at an angle of 23.5°.[4] As the earth goes around the sun, there are very subtle changes in both the shape of its orbit and in the tilt of its axis. For instance, the tilt of the earth's axis is decreasing very slowly. If one assumes that this motion has been going on for tens of thousands of years, then it would take about 41,000 years for the tilt to slowly go from 22.1° to 24.5° and back again. Since the seasons are caused by the tilt of the earth's axis, it should not be too surprising that uniformitarian scientists think that variations in the tilt could help contribute to an ice age.

Likewise, the earth's orbit around the sun is in the shape of a "squashed" circle called an *ellipse*. However, this ellipse is slowly becoming slightly less squashed. If one assumes that this subtle variation has been going on for many thousands of years, then this variation is characterized by cycles occurring over periods of about 100,000 and 405,000 years.

Precession is still another subtle motion in which the earth's axis slowly traces out a cone over about 26,000 years. Another kind of precession (*orbital precession*) causes the earth's elliptical orbit to slowly rotate relative to the background stars. These two precessions produce still another long cycle of about 22,000 years.

Mathematical equations can describe how these slow changes would presumably vary over long periods of time. Uniformitarian scientists "rewind" these equations back hundreds of thousands of years in order to determine the tilt of the earth's axis and the shape and orientation of its orbit at various times in the supposed "prehistoric" past (Figure 2). These changes would have caused the amount of summer sunlight falling on the mid-to-high northern latitudes to slowly increase and decrease over tens of thousands of years.

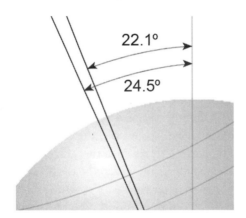

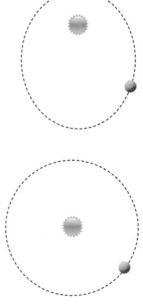

Figure 2. *Uniformitarian scientists believe that slow, subtle changes in the tilt of Earth's axis and the shape of its orbit around the sun varied the amount of high-latitude summer sunlight in the northern hemisphere over millions of years. Glacial intervals, or ice ages, are thought to result when this high-latitude summer sunlight is weakest.*

This is important because it is the summer months that determine whether or not an ice age can occur. In order for thick ice sheets to form, winter snow has to keep from melting during the summer, and this must be true for many years. Secular scientists generally believe that it is the amount of summer sunlight at 65° north latitude that "paces" the ice ages. This is because of the large amount of land at that latitude; since land masses cool much more quickly than the oceans, one would expect ice sheets to be most likely to form over land in the high polar latitudes.

Since snow is less likely to melt during cooler summers, secular scientists believe that ice ages occur at times when high-latitude summer sunlight is decreased. They then use mathematical equations to calculate the times in the "prehistoric" past when these decreases would have occurred. According to the astronomical theory, it is at these approximate times that the ice ages took place.

Right now this astronomical (or Milankovitch) theory is very popular among secular scientists. This is mainly because of a 1976 paper that showed apparently good agreement between the chemical "wiggles" in the seafloor sediments and the expectations of the astronomical theory.[5] But as we will later see, the astronomical theory has serious problems.

Determining Annual Layers in Ice Cores

Another factor in secularists' long-age calculations is the manner in which uniformitarian scientists date the Greenland and Antarctic ice sheets. Because snow generally does not completely melt at high altitudes and elevations, it accumulates over time. Thus, one layer of snow will be covered by a succeeding layer, that layer will be covered by still another layer, and so on. As the vertical thickness increases over time, the air is squeezed out and the snow transforms into ice.

This accumulated ice contains layers that can, in theory, be used to construct a chronology. For instance, summer and winter snow layers are generally distinct from one another. Likewise, variations in the acidity and dust content of the ice can also conceivably be used, due to seasonal fluctuations in these quantities.[6]

13

Secular scientists believe that other chemical clues in the ice can also yield information about past climates. For instance, the water (H_2O) in ice contains oxygen, so it is possible to calculate oxygen isotope ratio values for the ice and to measure how these $\delta^{18}O$ values vary with depth. Actually, creation scientists would agree that variations in the oxygen isotope ratio within the ice *can* yield clues about past climates, as interpretation of the chemistry of the ice is much less difficult than interpretation of the chemistry of fossils in seafloor sediments, although there are still some complicating factors that must be taken into account.[7]

As in the case of seafloor sediments, scientists drill and extract cores from the ice in the hopes that they can use layers within the cores to date events in Earth's past. Earth scientists would like to be able to determine the time that has elapsed since a layer within an ice core was deposited. One might think that this could be done by simply counting the annual layers within the ice, but in actual practice there are complicating factors.

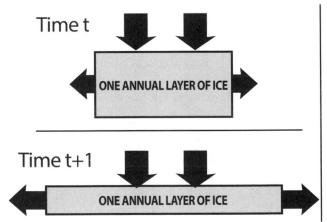

Figure 3. *Because layers of ice become thinner at increasing depths within an ice sheet, mathematical flow models must be used to determine how much thinning is present at a given depth.*

Although the simple counting of layers may often be feasible at the tops of the ice cores, it cannot consistently be extended to greater depths within the core. This is because seasonal layering becomes more indistinct at greater depths. Nor can one "guess" the locations of deeper layers based simply on the thicknesses of layers higher in the core. The weight of the overlying snow and ice causes the layers to be forced downward, with a corresponding thinning of the layers (Figure 3). This thinning is not constant as one goes deeper down into the core; rather, it increases with increasing depth within the core. Therefore, a mathematical flow model is needed to convert a measured length of the core into a time:

> In the upper part [of the cores], the timescale is established by counting annual layers in the ice, but eventually the layers become too thin to be identified and counted, and the timescale further down has to be established from models.[8]

Among other assumptions, these flow models implicitly assume that the ice sheets have been in existence for vast lengths of time.[9,10] Not surprisingly, they yield enormous ages for the ice sheets.

We are now ready (finally!) to discuss the connections between the dating of the high-latitude ice sheets, the dating of the seafloor sediments, and ice ages. Let's see why uniformitarian ice-age theories don't work.

2
THE ASTRONOMICAL THEORY
CANNOT EXPLAIN AN ICE AGE

Although uniformitarians believe many ice ages happened in the past, they have great difficulty explaining how or why even a single ice age would occur. Although the astronomical theory is currently popular, dozens of secular ice-age theories have been proposed over the years. However, *none* of these theories, including the astronomical theory, can adequately explain an ice age. In fact, one popular news magazine stated that the cause of ice ages was one of the "great mysteries" of science![1]

One big problem with the astronomical theory is that changes in high-latitude summer sunlight resulting from subtle changes in Earth's orbital motions should be small. These subtle variations in sunlight might act as a "pacemaker" to regulate the advance and retreat of ice sheets, but some other unknown mechanism is required to amplify these tiny changes so that ice ages can occur.

Incidentally, this is one of the reasons secular scientists tend to be much more concerned about global warming or climate change than creation scientists. Because secular scientists "know" that these subtle changes in sunlight "pace" the ice ages, and because they realize that the astronomical theory by itself is insufficient to cause ice ages, they generally assume that other unspecified mechanisms can amplify these subtle changes to bring about drastic climate change. This is why uniformitarian climatologists often mention "strong nonlinear responses" in their discussions of global warming or climate change. Therefore, they fear that human activity—such as the production of carbon dioxide gas from industry, etc.—could be just such an amplifying mechanism. So there is a subtle connection between acceptance of the uniformitarian astronomical theory and climate change alarmism.

One of the most serious problems with the astronomical theory is that the mechanism it proposes simply cannot, by itself, cause an ice age. However, there are other major problems:

- Because the tilt of the earth's axis has such a strong effect on our seasons, one would naturally expect that ice-age glacial-interglacial cycles should occur at 41,000-year intervals, corresponding to the expected time for the tilt to go from 22.1° to 24.5° and back again. Secular scientists claim that this was indeed the case between 2.6 million and 800,000 years ago. But after that, the length of the cycles somehow changed to about 100,000 years. Why would this happen?

- At first glance, these 100,000-year cycles may also seem reasonable, since one of the periods for changes in eccentricity (how "squashed" the earth's orbit is) is about 100,000 years. But of the astronomical cycles we have discussed, changes in eccentricity should have the *weakest* effect on climate. Why then would ice ages occur at 100,000-year cycles for the last 800,000 years?

Furthermore, based on a secular interpretation of the data, one could make a case that the second-to-last de-glaciation occurred about 10,000 years *before* the increases in high-latitude summer sunlight that would have caused the ice sheets to retreat (the "Termination II" problem)![2,3] The astronomical theory has many other problems, many of which are well-known to uniformitarian scientists and are even discussed within their own literature.[4.]

3

ASSIGNING LONG AGES TO ICE CORES INVOLVES CIRCULAR REASONING

If different dating methods really are telling such a consistent story about Earth history over millions of years, and these dating methods really are independent of one another, then this is indeed a very strong argument for a very old earth. But just how independent are the different dating methods? Is the dating of seafloor sediments and the ice cores somehow connected to uniformitarian ice age explanations?

In circular reasoning, an assumption is made and then is used to "prove" the assumption is true. Uniformitarian scientists routinely engage in this type of reasoning in their dating of ice cores and seafloor sediments.

Orbital Tuning

In recent years, uniformitarian scientists have become so convinced that the astronomical theory is correct that they use it to date the seafloor sediments. This technique is called *orbital tuning*.[1]

The orbital tuning method assumes that the "wiggles" in the chemistry of the seafloor sediments—such as changes in the oxygen isotope ratio ($\delta^{18}O$)—indicate changes in climate. These climate changes are then assumed to be caused by the astronomical cycles we discussed previously. Secular scientists run the astronomical calculations backward to calculate the approximate times at which ice ages would have occurred according to the astronomical theory. They then match the peaks in the $\delta^{18}O$ variable to these times. The orbital tuning method assumes that sediments containing high values of $\delta^{18}O$—which are thought to indicate ice ages—were deposited at the times the astronomical theory says the ice ages were supposed to have occurred!

Although uniformitarian scientists believe that the seafloor sediments were deposited slowly over millions of years, they do not believe that this rate of deposition was always constant. They believe that there were times when the sediments were deposited a little slower and times when they were deposited a little faster. So secular scientists assume whatever variable rates of deposition are needed to ensure that the peaks in these chemical "wiggles" will be deposited at the times in the past demanded by the astronomical theory. Because one can assume whatever past deposition rates are needed to get a match, even some secular scientist have criticized orbital tuning as a form of circular reasoning.[2] One specialist has noted:

> The possibility clearly exists to produce a tuned sedimentary series that has been forced to resemble an orbital template by overenthusiastic correlation.[3]

Moreover, secular scientists have not pinned down how long they think it would take the climate to respond to changes in the distribution of sunlight reaching the earth,[4] nor have they decided whether it is variations in summer sunlight at 65° north latitude or at some *other* latitude (say, 45° north) that are responsible for ice ages.[5] These uncertainties give them still more wiggle room (pun intended!) when using orbital tuning to date the seafloor sediments.

So uniformitarian scientists assume that the astronomical theory is correct (despite its problems), and they then use that assumption to date the seafloor sediments.

Of course, uniformitarian scientists recognize this potential for circular reasoning, and they attempt to guard against it by using what they would consider to be independent "checks" or constraints:

Orbital tuning is rarely applied to sediments without first

considering independent age constraints from fossil events and paleomagnetic reversals. These provide a preliminary age scale and therefore a guide to approximate, time-averaged, sedimentation rates to be modified by orbital tuning.[6]

But what are *paleomagnetic reversals* and *fossil events*, anyway?

Paleomagnetic Reversals

Paleomagnetic stratigraphy is a technique that secular scientists use to "tie" dates for the seafloor sediments to a record of past reversals in the earth's magnetic field. In this method, measurements of the weak magnetic signal found in iron-containing minerals within seafloor volcanic rocks is used to determine the direction of the earth's magnetic field at the time the rocks cooled.

Both creation and uniformitarian scientists generally agree that Earth's magnetic field has reversed multiple times in the past, with the north magnetic pole "trading places" with the south magnetic pole. Creation scientists believe that these magnetic reversals occurred very rapidly during the Flood cataclysm,[7] while uniformitarian scientists generally believe that these magnetic reversals occurred very slowly over thousands of years—in spite of the fact that secular scientists themselves have found evidence for *extremely* rapid magnetic reversals within volcanic rocks.[8] As new molten material comes up from the earth's interior at the mid-ocean ridges, the current orientation of Earth's magnetic field is recorded as the molten material cools. The boundaries between the positive and negative patterns in the volcanic rocks therefore indicate times at which the earth's magnetic field reversed or flipped.

Of course, the magnetic patterns do not themselves tell how long ago these magnetic reversals occurred. Uniformitarian scientists have relied on radioisotope dating methods, such as the potassium-argon (K/Ar) method, to assign ages to these magnetic reversals. The most recent of these reversals supposedly occurred about 780,000 years ago and is especially important for their dating of the seafloor sediments. Uniformitarian scientists see the time of this reversal as a chronological anchor or tie point, much in the same way that a biblical scholar would use important dates (such as dates for the Flood or the Exodus) as anchor points in the construction of a biblical chronology.

Creation scientists have shown that there are serious problems with radioisotope dating techniques and that the three main assumptions behind these methods are questionable.[9] Furthermore, the different dating methods are not truly independent of one another.

Fossil Events

Because secular scientists believe that sedimentary layers were deposited slowly over millions of years, they believe that the fossils within these layers provide a "snapshot" of life on Earth at a particular time in the past. Some fossils, called *index fossils*, have been found within relatively narrow ranges of sedimentary layers. Uniformitarian scientists interpret this to mean that these organisms lived within relatively brief intervals of time.

Suppose, for instance, that fossils of a particular creature have only been found within sedimentary strata that secular scientists date as being between 70 and 72 million years old. This age range may have been determined by radioisotope dating, not of the fossils or the sediments themselves, but of volcanic rocks above and below the sedimentary layers. Because uniformitarian scientists believe that this index fossil lived *only* between 70 and 72 million years ago, they believe they can use it to date other sedimentary rock layers. Should they find this same index fossil in another sedimentary layer, they will tend to date that layer as also being between 70 and 72 million years old. Although they believe that life has existed on Earth for billions of years, they assume that all fossils of this particular creature are the same age, regardless of where these fossils are found. In other words, they use the assumption that evolution is true in order to date the sedimentary rock layers! This is what evolutionary scientists mean when they speak of using *fossil events* or *faunal succession* to date the rocks.

However, there have been numerous instances in which index fossils have been found in layers *outside* of the range of strata in which they were previously found, clearly demonstrating that earlier attempts to use them for dating rocks were in error.[10] Furthermore, there have been past instances when dates resulting from fossil events have contradicted other dating methods, and the assumption of evolution trumped these methods.[11]

Of course, the Genesis Flood completely invalidates the fossil-events method. Since most fossil organisms were buried during this year-long global cataclysm, their locations within the sedimentary rocks tell us absolutely *nothing* about an alleged "prehistory" of millions of years. Their location within the sedimentary layers is simply the result of where they were buried during the Flood.

What Happens When Dating Methods Contradict the Theory?

As we have already noted, uniformitarian scientists use assumed dates for paleomagnetic reversals as chronological anchors for their dating of seafloor sediments. But the dates for these reversals are obtained using radioisotope dating techniques, such as the potassium-argon (K/Ar) method.

Uniformitarian scientists use these radioisotope methods to indirectly assist in their use of the astronomical theory to date the seafloor sediments. However, the radioisotope dates sometimes contradict the astronomical theory. One expert describes how a Dutch team of researchers concluded that previous K/Ar dates for magnetic reversal boundaries were a little younger than they "should" have been. Don't worry if you have trouble understanding the jargon in the first part of this excerpt; the important part has been highlighted:

> Hilgen and co-workers recognized orbital forcing by a grouping of sapropels (dark, organic-rich beds) into units of ~100 and 400 ky by eccentricity modulation of precessional climate changes. *Their resulting calibration of the GPTS* [geomagnetic polarity timescale] *yielded significantly greater ages for magnetic reversal boundaries than the previously accepted dates based on K/Ar radiometric age dating.*[12]

This is interesting for a number of reasons. Before the discovery of this "orbital forcing" pattern, the K/Ar dates had been assumed to be correct. But because they contradicted this newly discovered pattern, they were then assumed to be in need of "adjustment." Contrary to popular misconceptions, the dating methods *do* often contradict one another. Moreover, the astronomical theory is used as a filter that determines whether radioisotope dates are "good" or "bad": Good dates agree with the astronomical theory, and bad dates don't!

Secular scientists might argue that this adjustment of these particular radioisotope dates was justified because it was later "verified" by another radioisotope dating method:

> After initial controversy, the ages proposed by Hilgen and others have largely been verified by recent advances in $^{40}Ar/^{39}Ar$ dating of volcanic ash layers at a number of magnetic reversal boundaries.[13]

But is this really the case? In the $^{40}Ar/^{39}Ar$ (or Ar/Ar) method, a rock or mineral of unknown age and standard of "known" age are both bombarded with neutrons from a nuclear reactor in order to convert stable potassium-39 (^{39}K) into radioactive argon-39 (^{39}Ar). By determining the amount of ^{40}Ar in the rock compared to the radioactive ^{39}Ar resulting from this bombardment, and by making assumptions about how much of this ^{40}Ar resulted from the decay of ^{40}K, it is thought that the age of the rock may be calculated if the age of the standard is known. But how is the age of the standard determined?

One way is by using the K/Ar method.[14] So the Ar/Ar method is really just an extension of the K/Ar method. But secular scientists have another method at their disposal for calibrating the Ar/Ar method:

> The emerging astronomically calibrated geomagnetic polarity time scale (APTS) offers a means to calibrate the ages of $^{40}Ar/^{39}Ar$ dating standards that is independent of absolute isotopic abundance measurements.[15]

So secular scientists also use the astronomical theory to calibrate standards used in the Ar/Ar method. Who would have ever guessed? And don't forget that it was the apparent agreement between the astronomical theory and the Ar/Ar method that supposedly justified the rejection of the older K/Ar dates in favor of the newer Ar/Ar dates in the first place!

Note the circular reasoning involved in this entire process. Secular scientists use one radioisotope dating method to calibrate another method. They then use the radioisotope dating methods to assist them as they use the astronomical theory to date the seafloor sediments. But they then use the astronomical theory to (1) judge between "good" and "bad" radioisotope dates and (2) to calibrate the radioisotope dating methods!

Creation researcher Marvin Lubenow astutely observed, "In the dating game, evolution always wins."[16] If a date contradicts the evolutionary story, one can always find an excuse for changing it or getting rid of it. Many examples of this kind of circular reasoning can be cited since it is rampant in the historical sciences.[17,18]

Despite popular misconceptions, the different dating methods are not independent of one another. They generally agree with one another because they are being *made* to agree.

Dating of the Greenland and Antarctic Ice Cores

While it is true that uniformitarian scientists do count visible layers (more on that later) as they attempt to date the ice cores, these counting methods are not the most common dating method:

> This property [varying strain rate for ice] has important implications for ice-sheet modelling, *the most frequently used method for dating ice cores*.[19]

Because uniformitarian flow models assume that the ice sheets have been in existence for millions of years, the fact that they yield old ages for the ice cores is hardly surprising. As in the case of the seafloor sediments, uniformitarian scientists attempt to use "independent" dating methods to constrain their flow models:

> One dimensional (1-D) ice flow models are used to construct the age scales at the Dome C and Dome Fuji drilling sites (East Antarctica). The poorly constrained glaciological parameters at each site are recovered by fitting independent age markers identified within each core.[20]

As one might suspect, however, these "independent age markers" are not truly independent. For instance, researchers describe how they used orbital tuning to date an ice core that had been drilled at Vostok station in Antarctica:

> We have applied the orbital tuning approach to derive an age-depth relation for the Vostok ice core, which is consistent with the SPECMAP marine time scale.[21]

The researchers noted that their age-scale agreed with the SPECMAP marine timescale, a chronology that had been constructed for the seafloor sediments. But this agreement is hardly surprising, since the SPECMAP marine timescale was *also* obtained by orbital tuning![22] The authors also used a second method to date the Vostok core:

> A second age-depth relation for Vostok was obtained by correlating the ice isotope content with estimates of sea surface temperature from Southern Ocean core MD 88-770.[23]

Therefore, the dating of the Vostok ice core was not independent of other methods. Rather, orbital tuning and seafloor sediments were used. This certainly comes as a surprise to many—who would have guessed that secular scientists would be using seafloor sediments in order to date layers within the Antarctic ice sheet?

Moreover, secular scientists have used ice cores to help them date the seafloor sediment cores. This was the case for the MD97-2120 sediment core that was obtained from the Chatham Rise east of New Zealand. The tie points used to construct the age model for the lower portion of this core were obtained by correlating assumed sea-surface temperatures to variations in the chemistry of Antarctica's Vostok ice core. But these chemical variations had themselves been orbitally tuned! Likewise, the chronology of a middle section of the core was obtained by "tuning" to $\delta^{18}O$ variations within another deep-sea core, the MD95-2042 core—which is located off the coast of Portugal. But the chronology of the MD95-2042 sediment core was *itself* tied to an ice core from Greenland![24]

In summary, secular scientists use the assumption that evolution and the astronomical theory are correct to date the seafloor sediments. Paleomagnetism is also used to assist in this process, but radioisotope dating methods are required to date the times of these magnetic reversals. However, the astronomical theory is used to calibrate dates obtained by radioisotope dating methods. Finally, the seafloor sediments are used to help date other sediment cores. These sediment cores are used to set the timescales for the glacial-flow models, which are then used to date the ice cores (Figure 4). Then, coming full circle, secular scientists have even used ice cores to date the seafloor sediments!

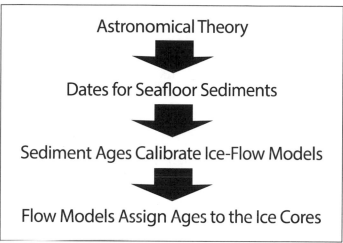

Figure 4. *Secular scientists assume the astronomical theory of ice ages is correct, despite its problems. The astronomical theory then assigns dates to the seafloor sediments. The dates for the seafloor sediments are then used to calibrate secular ice-flow models, which in turn are used to date the ice cores.*

Notice that the assumption of uniformitarianism, the philosophy about which the apostle Peter warned us, undergirded this entire process. Because secular scientists deny the Bible's short timescale, they believe that the heavenly bodies have existed for billions of years and that the high-latitude ice sheets have existed for millions of years. Because they deny the Flood, they assume that deposition of seafloor sediments has always been slow and gradual. They then point to the apparent (but forced) agreement between their uniformitarian philosophy and their interpretation of the seafloor sediments and ice cores, and then they use this agreement to browbeat Christians, demanding that Christians abandon a straightforward understanding of Scripture!

In other words, they are assuming that uniformitarianism is true, using that assumption to make their interpretation of the data agree with their philosophy, and then using this agreement as "proof" that their uniformitarian philosophy—which includes evolution and millions of years—is correct. In essence, it is a gigantic exercise in circular reasoning.

4
OTHER METHODS USED IN DATING THE ICE CORES

While believers in an old earth might acknowledge that their theoretical ice-flow models do indeed assume an old earth, they might argue that this assumption is justified since glaciologists also use other methods to date the ice cores. These other dating methods include visual counting of what are assumed to be annual patterns within the ice cores.

For instance, visual identification of what are called *depth hoar complexes* is often used in the upper core segments. Depth hoar is low-density snow characterized by large, cup-shaped ice crystals. Depth hoar can form in clear, calm weather when the temperature above the snow changes rapidly with increasing height. If this clear weather is followed by a large storm, a crisp surface called a *wind crust* or *wind slab* can form above the depth hoar. Such conditions can occur repeatedly during the summer months, forming a depth hoar complex.[1]

Other dating methods include the counting of presumed seasonal variations in the chemistry of the ice (including variations in the oxygen isotope ratio, $\delta^{18}O$) and presumed seasonal variations in the dust content and electrical conductivity of the ice.

Likewise, ice cores can contain debris (tephra) from volcanic eruptions, as well as layers containing greater amounts of sulfates resulting from volcanic eruptions. Since the dates of powerful volcanic eruptions are thought to be known, secular scientists believe that they can use these *volcanic reference horizons* as "checks" on their counting methods and flow models.

At first glance, it might seem reasonable to think that these methods can prove that the ice sheets in Greenland and Antarctica have been in existence for extremely long periods of time. But in actual practice things are more complicated. For instance, how can one be certain that a particular layer is annual rather than sub-annual? Even secular scientists have acknowledged this difficulty:

> We certainly must entertain the possibility of misidentifying the deposit of a large storm or a snow dune as an entire year or missing a weak indication of a summer and thus picking a 2-year interval as 1 year.[2]

Likewise, layer-counting methods can be subject to large errors.

> In common with all other methods of layer counting, visible stratigraphy has certain weaknesses. Probably the biggest weakness is that the signal is variable along the full core length (sometimes subtle, sometimes obvious); unless the observer is careful and persistent, it is likely that large errors will result.[3]

Furthermore, some of the more common layer-counting methods have the following weaknesses.

Depth Hoar Complexes

Multiple depth hoar/wind crust patterns can form in a single year, and scientists have repeatedly observed 15 to 16 storms per year in central Greenland.[4] Even secular scientists have acknowledged that two depth hoar events that formed during the same year could be mistaken for two separate "years" if they were physically separated by a significant depth of snow or ice.[5] These depth hoar complex patterns are only evident in the upper portions of the Greenland ice cores, so other methods must be used to date the deeper portions of these cores.

The small amounts of snowfall in central Antarctica, together with the blowing wind, preclude the use of these patterns to date the deep cores from central Antarctica. Therefore, the dating of these cores is especially dependent upon theoretical ice-flow models.

Although the coastal regions of Antarctica receive greater amounts of snowfall, these coastal Antarctic cores are generally not as deep, and the ages assigned to them are much younger.

Dust Layer Counts

The deeper parts of the ice cores contain higher levels of dust. This is especially true of the Greenland cores, which typically have average dust concentrations that are about 12 times greater than in the upper portions of the cores. But this dust content can vary greatly even in the bottom parts of the cores, with dust levels ranging anywhere from 3 to 70—and even *100*—times greater than dust levels in the upper portions![6,7] Within the Greenland GISP2 core, for instance, the depth hoar complexes in the upper half of the core were replaced by cloudy, dust-laden bands that were assumed to indicate seasonal variations in dust content. But dust levels can increase due to short-term factors like dust storms and volcanic eruptions, and they can vary significantly over even short distances.[8]

Counting of Seasonal Chemical "Wiggles"

Although seasonal "wiggles" in the chemistry of the ice (including variations in $\delta^{18}O$) are evident in the upper portions of the cores, they become indistinct at greater depths. For instance, seasonal $\delta^{18}O$ fluctuations within the Greenland GISP2 ice core could not be clearly identified at depths greater than 300 meters.[9] Moreover, how does one unambiguously determine whether a "wiggle" in the measured $\delta^{18}O$ indicates a seasonal variation or one of shorter duration?

Variations in Electrical Conductivity

The acidity of snow and ice is generally greater during the spring/summer months. The presence of these acids makes it easier for electricity to pass through the ice, corresponding to an increase in the ice's electrical conductivity. However, this method is generally of limited use in the bottom portions of the cores since their higher dust concentrations dramatically reduce the electrical conductivity of the ice. Moreover, it should be remembered that multiple acid peaks can be caused by other factors (such as volcanic eruptions), and multiple acid peaks have been observed to form within a single year.[10]

Volcanic Reference Horizons

Unfortunately, accurate historical dates (recorded by eyewitnesses) for large volcanic eruptions are generally only known for the last 300 years,[11] with a few large volcanic eruptions that can potentially be historically dated as far back as 2,000 years ago.[12] And it is the dating of the *lower* parts of the cores that is controversial. Therefore, this method cannot really serve as a check on the dating of these lower portions, with their presumed timescales of hundreds of thousands of years. Secular scientists claim that radioisotope dating can be used to date older volcanic eruptions, but, as noted earlier, there are serious problems with radioisotope dating methods.

The GISP2 Greenland Ice Core: The "Ultimate Proof" against Noah's Flood?

It has been claimed that the GISP2 ice core is the "ultimate proof" that Noah's Flood could not have happened as described in the Bible.[13] This is because secular scientists claim to have counted 110,000 annual layers within the ice down to a depth of 2,800 meters. Based on the Bible's geological and chronological information, biblical scholars believe that Noah's Flood occurred about 4,500 years ago. Moreover, any prior ice sheets that might have existed would surely have been destroyed during the Flood cataclysm. If Genesis is true, the Greenland ice sheet has only had about 4,500 years to form. How then can it contain more than 100,000 annual layers? Therefore, it is claimed that Noah's Flood could not have been the worldwide, cataclysmic event described in the Bible.

Moreover, the dating of much of the GISP2 core relied heavily on visual identification of layers.[14] If scientists were able to actually *see and count* a total of 110,000 annual layers, this would indeed be a powerful argument against the Genesis Flood, as well as a powerful argument for an old earth. How do creation scientists respond to this claim?

For one thing, thick ice sheets simply do not need vast amounts of time to form. For instance, Crater Glacier on Mount St. Helens has an average thickness of 300 feet (and is more than 600 feet thick in some locations), yet it is less than 40 years old. Thus, it is reasonable for both the

Jake Hebert

Greenland and Antarctic ice sheets to have also formed rapidly. Likewise, one can obtain a ballpark estimate of the times required for the Greenland and Antarctic ice sheets to form. Even if one ignores the fact that snowfall would have been much greater during the Ice Age, the Greenland ice sheet would only need 5,000 years to form, and the Antarctic ice sheet would only require 10,200 years (assuming no melting of the ice).[15]

Also, layer-counting methods are subject to the weaknesses we have already noted, and bias plays a large role in whether or not a layer will be considered annual. This was dramatically illustrated in the dating of the GISP2 core.

As indicated earlier, the GISP2 scientists counted depth hoar complexes to date the upper 1,500 meters of the core. Secular scientists believe that the GISP2 ice at a depth of 1,500 meters is about 9,300 years old, and the fact that multiple depth hoar/wind crust patterns can form within a single year makes it fairly easy to account for the 5,000 or so "extra" years in the upper core.[16,17] But what about the extreme ages assigned to the bottom part of the core?

Below 1,500 meters the scientists counted cloudy, dust-laden bands that were thought to represent seasonal variations in dust content. However, they sometimes had difficulty visually discerning these dust bands, and this was especially true in a 500 meter-long section in the deepest part of the core. For this reason, they were especially dependent on a technique called *laser light scattering* (LLS) to identify presumed annual peaks in dust content. In the LLS method, the amount of laser light reflected from either a sample of melted ice or the ice core itself is used as an indicator of the amount of dust within the ice.

In their original attempt to date the bottom of the core using the LLS method, the scientists used a laser beam with a diameter of 8 millimeters. They concluded that the ice at a depth of 2,800 meters was about 85,000 years old.

However, they realized that this result contradicted another chronology that suggested that the ice at that depth should be about 110,000 years old. One described the problem:

[Other scientists] predicted the age of the ice at 2800 m to be about 110,000 years, 25,000 years older than had been originally counted on the basis of visual stratigraphy....

Because of the above discrepancy the senior author returned to the National Ice Core Laboratory and rechecked the visible stratigraphy. No significant changes from the original counts were observed. Study of the LLS record, now available in a more detailed format using 1-mm beam width rather than 8 mm..., showed there was more structure indicating a greater number of annual layers, leading to a more expanded timescale than had been interpreted from the stratigraphic record.[18]

After two scientists re-counted dust peaks in this 500 meter-long section using a much narrower (1 mm) beam, they averaged their two counts, obtaining a number that agreed with the expected result to within about 1%:

This average was then compared to the Sowers-Bender correlated timescale and showed a maximum difference of 1.1% with an age of approximately 111,000 B. P. [before present].[19]

But the only reason these scientists bothered re-counting the dust peaks in the first place was because they had gotten the "wrong" answer when they first counted them. They needed an extra 25,000 dust layers in order for their result to agree with the "correct" answer...and they found them. Clearly, the counting of "annual" layers within ice cores is a highly subjective process!

This subjectivity is highlighted still further by the fact that one of the scientists consistently counted 20% more dust layers than did the other.[20]

It should be noted that more than 67,000 of these alleged 110,000 annual layers were counted in just this 500 meter-long section!

But even with the subjectivity and circular reasoning involved in these dating methods, there are still discrepancies that need to be resolved, and not just in the lower parts of the core but also in the *upper* sections—although these contradictions are not well-known by the general public.[21]

Jake Hebert

Even "Simple" Counting Is Influenced by Beliefs about the Past

The above example illustrates that even "simple" counting of layers is not really independent of one's beliefs about the past. When using the laser light scattering (LLS) method, the GISP2 scientists needed four to five measurements in order to discern a peak in dust content.[22] But how far apart should they make those measurements? At a given depth in the ice, what length of the ice core corresponds to a year?

Well, that depends upon how thin one believes an annual layer at that depth will be. If you believe that annual layers at a given location in the ice are very thin, as expected from the secular ice-flow models, then you will expect these annual dust peaks to be very close together. So in order to clearly "see" these peaks, you must make a very large number of closely spaced measurements. Likewise, you should use a smaller-diameter laser beam in order to make sure that you don't overlook small fluctuations in dust content that might be thinner than the width of your laser beam.

But because dust levels are never perfectly constant, there will always be variations in dust content, no matter how closely spaced your measurements are. Unless you already know in advance the thickness of an annual layer of ice at a given depth, there is *no way to know* whether or not these fluctuations in dust levels are annual or sub-annual variations. And remember that dust concentrations can vary dramatically in the bottom sections of the Greenland cores, which greatly increases the likelihood that sub-annual variations in dust content will be mistaken for annual ones.

Thus, it is the theoretical ice-flow model chosen by the scientist that ultimately determines the expected annual thickness of the ice at a given depth, and it is this expected thickness that determines both the diameter of the laser beam used and the spacing of the measurements made. If you are *expecting* to find a great number of "annual" dust peaks, you will find them because dust content will inevitably vary with depth in the ice regardless of whether your measurements are separated by meters, centimeters, or millimeters. In fact, this is true for any quantity measured at different depths within the ice.

So even though "simple" counting of layers may superficially *seem* to be independent of one's beliefs about Earth history, it really isn't. Old-earth beliefs subtly guide the counting process to ensure that the results agree with old-earth beliefs!

5

THE BIBLE EXPLAINS
THE ICE AGE

We have seen that secular ice-age explanations are inadequate and that secular scientists regularly engage in circular reasoning when dating the ice cores. Why is an ice age so hard for secular scientists to explain?

Naively, one might think that colder winters would be sufficient to produce an ice age, but this is not the case—very cold temperatures generally result in *less* snowfall due to the lower moisture content of very cold air. Also, there are places on Earth today that experience very cold winters but are not characterized by glaciers because the warm summers melt the winter snow before glaciers can form.

Cold summers and much more snowfall are needed for an ice age, and these conditions must continue for many years so that snow and ice can accumulate. Today, it is very difficult to meet both these conditions for any extended period of time. Realistic computer simulations have shown that even a dramatic 12°C summer temperature drop in northeastern Canada would produce only a modest advance in permanent snow cover; the snow cover would not even extend past the southern tip of Canada's Hudson Bay.[1] Yet the geological evidence indicates that glaciers once extended well into the northern United States. Past conditions must have been radically different in order to produce an ice age.

The Real Cause of the Ice Age, in a Nutshell

The worldwide Genesis Flood in the days of Noah, however, would have provided these conditions. One can use the acronym HEAT to remember the four key points of this explanation.

Hot oceans. During the Genesis Flood, hot, molten material from Earth's interior (possibly including much warmer waters from the "fountains of the great deep," Genesis 7:11), volcanism, and friction from plate tectonics would have significantly warmed the world's oceans, perhaps by tens of degrees Celsius.

Evaporation. Warmer oceans would have resulted in greatly elevated evaporation. This would have increased the amount of moisture in the atmosphere, ultimately resulting in much greater snowfall over the relatively cool continents in the mid- and high-latitude regions.

Aerosols. The enormous amounts of volcanic activity that occurred toward the end of the Flood and afterward would have put an enormous volume of ash and tiny particles called *aerosols* into the atmosphere. These aerosols would have reflected significant amounts of sunlight away from Earth's surface, resulting in cooler summers over the continents. Thus, winter snow and ice would not completely melt, even during the warmest months. Ice sheets would grow as more snow and ice accumulated during subsequent winters.

Time. Explosive volcanic eruptions can result in noticeable cooling over the continents, and both creation and evolution scientists agree that many enormous volcanic eruptions have occurred in the past. Creation scientists believe many of these eruptions occurred toward the end of the Flood and for many years afterward as Earth slowly returned to equilibrium. Aerosols from explosive volcanic eruptions are a potent cooling mechanism to keep developing ice sheets from melting. However, because uniformitarian scientists claim that thousands, and even millions, of years separated these large volcanic eruptions, they cannot use this mechanism to account for an ice age. Thus, the Bible's short timescale is *critical* in explaining the Ice Age. (These points are summarized in Figure 5.)

Let's walk through these key points one by one.

Hot Oceans

While the notion that warmer oceans would result in greater evaporation is not controversial, how does one warm the world's oceans by tens of degrees Celsius? It would take an enormous amount of energy to raise their temperatures even *one* degree Celsius.

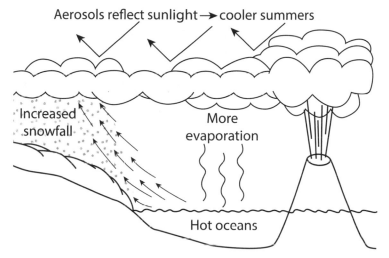

Figure 5. *Hot oceans after the Genesis Flood resulted in much greater evaporation. The increased moisture in the atmosphere then resulted in greater snowfall at higher latitudes and elevations. Residual volcanic activity after the Flood cataclysm increased atmospheric aerosols, which reflected large amounts of sunlight, providing the necessary summer cooling to keep the snow from melting so that thick ice sheets could form. The Bible's short timescale prevents this cooling effect from being "diluted" by long periods of time between the eruptions.*

Such a large change in ocean temperature would require a truly dramatic event. Uniformitarian scientists, who deny that the Flood of Noah occurred, cannot even consider such an increase in ocean temperature as a cause for an ice age.

However, the Bible says that the Flood catastrophe began when "all the fountains of the great deep were broken up" (Genesis 7:11). These fountains, which could have included the pre-Flood oceans as well as large amounts of water located below the earth's crust, could have been the source for most, if not all, of the waters of the Flood. This implies large-scale volcanic and tectonic activity, demonstrating that the Genesis Flood was a violent, catastrophic event. Creation scientists believe there was a complete turnover of the seafloor during the Flood, creating an entirely new ocean crust composed of freshly extruded lava. The heat resulting from this volcanic activity, and possibly tectonic friction and meteoritic impacts, would have contributed to the warming of the post-Flood oceans. If these fountains did indeed include large subterranean amounts of water, they too would have contributed to the warming, since these waters would have likely been quite warm—temperatures within the earth's crust typically increase by about 25°C per kilometer of depth. Flood turbulence and currents and earth movements would have mixed the water so that after the Flood the oceans would have been very uniform in temperature (unlike today's oceans).

Evaporation

Higher sea-surface temperatures result in greater evaporation. Therefore, much warmer oceans could dramatically increase the amount of moisture in the atmosphere. This increased atmospheric moisture would have resulted in much greater precipitation (rainfall and snowfall). Snow would have fallen at high latitudes and elevations.

Prior to his retirement, Dr. Larry Vardiman, an atmospheric scientist at ICR, and Dr. Wes Brewer, a computer expert, used a standard meteorological computer program to simulate the effects of significantly higher ocean temperatures on precipitation rates. After calibrating the program to ensure that the program accurately modeled precipitation rates for known meteorological events, the program was used to simulate precipitation resulting from higher ocean temperatures. Of particular interest was the effect of higher Pacific Ocean sea-surface temperatures on snowfall in California's Yosemite National Park, since Yosemite was once covered by thick glaciers.[2] Could such glaciers form within the biblical timeframe?

The simulation results showed that significantly warmer ocean temperatures could result in dramatic increases in snowfall, perhaps four to eight times greater than what occurs today.[3] Vardiman and Brewer also did similar simulations for Yellowstone National Park.[4] These simulations showed that glaciers more than a kilometer thick could easily have developed in just a few hundred years after the Genesis Flood.

Aerosols

Both creation and evolution scientists agree that massive amounts of volcanic activity have occurred in Earth's past. Some of these eruptions were explosive in that they ejected large amounts of dust and ash into the atmosphere. In addition to the dust and ash, large amounts of sulfur-containing gases were ejected. Through atmospheric chemical reactions, this sulfur resulted in the production of tiny sulfuric acid droplets called *aerosols*. These aerosols—which can remain in the atmosphere for two to three years—can reflect significant amounts of sunlight back into space. This reduced sunlight results in a noticeable cooling effect that has been observed to be greatest in the summer and autumn months.[5] Within recent history, such a cooling effect has been observed after large explosive volcanic eruptions, including the 1783 Laki eruption in Iceland, the 1815 eruption of Tambora, the 1883 eruption of Krakatoa, and the 1991 eruption of Pinatubo.

However, the volcanic eruptions of the past were vastly larger than today's eruptions. The well-known 1980 eruption of Mount St. Helens ejected about a cubic kilometer of ash into the atmosphere. In fact, compared to past volcanoes, the Mount St. Helens eruption was fairly small, resulting in a global cooling effect of only about a tenth of a degree Celsius, due to a smaller amount of ejected sulfur. Other explosive eruptions in the recent past have resulted in larger decreases in temperature. After the 1883 eruption of Krakatoa, average global temperatures dropped by as much as 1.2°C.

However, geologists have deduced that explosive eruptions in the distant past ejected *thousands* of times more ash into the atmosphere,[6] and these larger volumes of ash would presumably have been accompanied by even larger amounts of sulfur-rich gases, which would have resulted in even greater reductions in global temperature.

The large amounts of volcanic activity during and after the Flood would have provided the cooling needed to keep winter snow from melting during the summer months. In following winters, still more snow would accumulate, and the snow would be transformed into ice. This resulted in the formation of the large high-latitude ice sheets in the northern and southern hemispheres.

It is worth noting that many hundreds of reference horizons from explosive volcanic eruptions have been identified in the lower Ice Age portion of the GISP2 core.[7]

Time

Although this cooling mechanism may seem plausible, there is a potential problem with it. As noted earlier, aerosols fall out of the atmosphere within a few years. This cooling effect will therefore be fairly short-lived, even after extremely large volcanic eruptions. Yet these cooler summers must persist for *many* years if the ice sheets are to grow. Is this a problem for the biblical Ice Age model?

No, it is not. After the Flood, tectonic and volcanic activity did not come to an abrupt halt, since the earth was still reeling from the Flood cataclysm. Even within the secular model, "recent" Earth history (the last five million years or so) was a time of continued mountain building and volcanic activity. In the creation model, much of this mountain building occurred in the latter stages of the Flood and in the years following. Thus, volcanic activity would have continued intermittently after the Flood, with the intensity of these eruptions gradually decreasing with time. Today's relatively small volcanic eruptions are a faint "echo" of the enormous volcanic activity that occurred during and after the Flood. These continuing volcanic eruptions would have replenished the atmospheric aerosols, ensuring cool summers for many years.

The fact that explosive volcano eruptions can cause global cooling is well known, and secular scientists are well aware that past volcanic eruptions were *much* larger than those of today. Why then is the Ice Age still a mystery to secular scientists?

Remember that uniformitarian scientists believe that these enormous volcanic eruptions were separated by thousands, and even millions, of years. Although noticeable cooling would be expected to follow these eruptions, the cooling would not last long enough for thick ice sheets to develop. In order for thick ice sheets to form, these volcanic eruptions must occur close together in time.

Therefore, the Ice Age is still a mystery to uniformitarian scientists because of their belief in millions of years! Since they incorrectly believe these volcanic eruptions to be separated by vast amounts of time, they are unable to fully make use of this potent cooling mechanism.

Thus, we see that the Bible's short 6,000-year timescale, rather than being an impediment to scientific understanding, is actually one of the keys that enable us to explain the Ice Age.

CONCLUSION

In the creation model, the post-Flood Ice Age would have been a relatively short event. As volcanic activity decreased and the oceans gradually cooled, the Ice Age would have eventually come to an end. The end of the Ice Age would have been associated with much colder, drier conditions that would have led to the extinction of many large animals such as the wooly mammoth.

Much more can be said on these topics, so I encourage you to take advantage of the wealth of information about the Ice Age found on the Institute for Creation Research website—especially at icr.org/ice-age—as well as on the websites of other creation ministries such as Answers in Genesis and Creation Ministries International. The amount of research that creation scientists have done on the topics of the Ice Age, the ice cores, and the circular nature of uniformitarian dating methods simply cannot all be contained in this one little booklet.

Proponents of evolution and an old earth cannot adequately explain the Ice Age, the most recent major event in Earth history. Why then would Christians trust their claims about events that supposedly occurred hundreds of millions, and even billions, of years ago? As believers, we can look at the data and confidently embrace the discoveries of science, knowing that the facts consistently confirm what we read in God's Word. In science, as in life, the Bible holds the key.

NOTES

Introduction

1. Walker, M. and J. Lowe. 2007. Quaternary science 2007: a 50-year retrospective. *Journal of the Geological Society.* 164 (6): 1073–1092.
2. Meese, D. A. et al. 1997. The Greenland Ice Sheet Project 2 depth-age scale: Methods and results. *Journal of Geophysical Research.* 102 (C12): 26411–26423.
3. Johnson, J. J. S. 2013. Genesis Data Add Up to a Young Earth. In *Creation Basics & Beyond: An In-Depth Look at Science, Origins, and Evolution.* Dallas, TX: Institute for Creation Research, 47–53.
4. In order to present the maximum possible amount of information, I have not provided references for statements that are generally non-controversial or can easily be verified by the reader. Rather, I have reserved references for statements that are either less well-known or are more likely to be challenged by secular critics of the biblical creation-Flood model. Many of the sources referenced may be read for free on the Internet.

Chapter 1

1. There are very good reasons to believe that most of the seafloor sediments were deposited very rapidly. See Patrick, K. 2010. Manganese nodules and the age of the ocean floor. *Journal of Creation.* 24 (3): 82–86.
2. Wright, J. D. 2010. Cenozoic Climate–Oxygen Isotope Evidence. In *Climate and Oceans.* Steele, J. H., S. A. Thorpe and K. K. Turekian, eds. Amsterdam: Academic Press, 316-327.
3. Oard, M. J. 1984. Ice Ages: The Mystery Solved? Part II: The Manipulation of Deep-Sea Cores. *Creation Research Society Quarterly.* 21 (3): 125–137.
4. This tilt is measured from a line that is perpendicular to the imaginary plane containing the earth's orbit around the sun.
5. Hays, J. D., J. Imbrie and N. J. Shackleton. 1976. Variations in the Earth's Orbit: Pacemaker of the Ice Ages. *Science.* 194 (4270): 1121–1132.
6. Meese, D. A. et al. 1997. The Greenland Ice Sheet Project 2 depth-age scale: Methods and results. *Journal of Geophysical Research.* 102 (C12): 26411–26423.
7. Vardiman, L. 2001. *Climates Before and After the Genesis Flood.* El Cajon, CA: Institute for Creation Research, 41–68.
8. Using ice flow models for dating. Niels Bohr Institute: Centre for Ice and Climate. Posted on iceandclimate.nbi.ku.dk.
9. Cuffey, K. M. and W. S. B. Paterson. 2010. *The Physics of Glaciers,* 4th ed. Amsterdam: Butterworth-Heinemann, 569–574.
10. Ice sheet models. Niels Bohr Institute: Centre for Ice and Climate. Posted on iceandclimate.nbi.ku.dk.

Chapter 2

1. Watson, T. 1997. What causes ice ages? *U.S. News & World Report.* 123 (7): 58–60.
2. Oard, M. J. 1999. Another threat to the Milankovitch theory quelled? *Journal of Creation.* 13 (1): 11–13.
3. Shakun, J. D. et al. 2011. Milankovitch-paced Termination II in a Nevada speleothem? *Geophysical Research Letters.* 38 (18): L18701.
4. Cronin, T. M. 2010. *Paleoclimates: Understanding Climate Change Past and Present.* Columbia University Press, New York: 130–139.

Chapter 3

1. Herbert, T. D. 2010. Paleoceanography: Orbitally Tuned Timescales. In *Climate and Oceans.* Steele, J. H., S. A. Thorpe and K. K. Turekian, eds. Amsterdam: Academic Press, 370–377.

2. Ibid, 374.
3. Ibid, 372.
4. Ibid.
5. Cronin, T. M. 2010. *Paleoclimates: Understanding Climate Change Past and Present.* Columbia University Press, New York: 119.
6. Herbert, Paleoceanography: Orbitally Tuned Timescales, 373.
7. Humphreys, D. R. 1990. Physical Mechanisms for Reversals of the Earth's Magnetic Field During the Flood. In *Proceedings of the Second International Conference on Creationism*, vol. II. Walsh, R. E., ed. Pittsburgh, PA: Creation Science Fellowship, 129–142.
8. Snelling, A. 1995. The 'principle of least astonishment'! *Journal of Creation.* 9 (2): 138–139.
9. Vardiman, L. et al. 2003. Radioisotopes and the Age of the Earth. In *Proceedings of the Fifth International Conference on Creationism.* Ivey, Jr., R. L., ed. Pittsburgh, PA: Creation Science Fellowship, 337–348.
10. Oard, M. 2000. How well do paleontologists know fossil distributions? *Journal of Creation.* 14 (1): 7–8.
11. Lubenow, M. L. 1995. The pigs took it all. *Creation.* 17 (3): 36–38.
12. Herbert, Paleoceanography: Orbitally Tuned Timescales, 374, emphasis added.
13. Ibid.
14. Some problems with the $^{40}Ar/^{39}Ar$ technique: Standard Intercalibration. New Mexico Geochronology Research Laboratory K/Ar and $^{40}Ar/^{39}Ar$ Methods. New Mexico Bureau of Geology and Mineral Resources. Posted on geoinfo.nmt.edu .
15. Renne, P. R. et al. 1994. Intercalibration of astronomical and radioisotopic time. *Geology.* 22 (9): 783–786.
16. Lubenow, The pigs took it all.
17. Oard, M. J. 2001. End-Mesozoic extinction of dinosaurs partly based on circular reasoning. *Journal of Creation.* 15 (2): 6–7.
18. Oard, M. J. 2013. The reinforcement syndrome ubiquitous in the earth sciences. *Journal of Creation.* 27 (3): 13–16.
19. Paterson, W. S. B. 1991. Why ice-age ice is sometimes "soft." *Cold Regions Science and Technology.* 20 (1): 75–98, emphasis added.
20. Parrenin, F. et al. 2007. 1-D-ice flow modelling at EPICA Dome C and Dome Fuji, East Antarctica. *Climate of the Past.* 3 (2): 243–259.
21. Waelbroeck, C. et al. 1995. A comparison of the Vostok ice deuterium record and series from Southern Ocean core MD 88-770 over the last two glacial cycles. *Climate Dynamics.* 12 (2): 113–123.
22. Gornitz, V., ed. 2009. SPECMAP. In *Encyclopedia of Paleoclimatology and Ancient Environments.* Dordrecht, NL: Springer, 911.
23. Waelbroeck, C. A comparison of the Vostok ice deuterium record and series from Southern Ocean core MD 88-770.
24. Pahnke, K. et al. 2003. 340,000-Year Centennial-Scale Marine Record of Southern Hemisphere Climatic Oscillation. *Science.* 301 (5635): 948–952. A summary of the methods used to date this sediment core was archived (as of May 15, 2014) on the National Oceanic and Atmospheric Administration National Climatic Data Center website at ncdc.noaa.gov.

Chapter 4

1. Alley, R. B. 1988. Concerning the Deposition and Diagenesis of Strata in Polar Firn. *Journal of Glaciology.* 34 (118): 283–290.
2. Alley, R. B. et al. 1997. Visual-stratigraphic dating of the GISP2 ice core: Basis, reproducibility, and application. *Journal of Geophysical Research.* 102 (C12): 26378.
3. Ibid, 26377.
4. Alley, R. B. and B. R Koci. 1988. Ice-Core Analysis at Site A, Greenland: Preliminary Re-

sults. *Annals of Glaciology.* 10: 1–4.

5. Shuman, C. A. and R. B. Alley. 1993. Spatial and temporal characterization of hoar formation in central Greenland using SSM/I brightness temperatures. *Geophysical Research Letters.* 20 (23): 2643–2646.

6. Paterson, W. S. B. 1991. Why ice-age ice is sometimes "soft." *Cold Regions Science and Technology.* 20 (1): 75–98.

7. Ruth, U. et al. 2003. Continuous record of microparticle concentration and size distribution in the central Greenland NGRIP ice core during the last glacial period. *Journal of Geophysical Research.* 108 (D3): 4098.

8. Ram, M. et al. 1995. Polar ice stratigraphy form laser-light scattering: Scattering from ice. *Geophysical Research Letters.* 22 (24): 3525–3527.

9. Meese, D. A. et al. 1997. The Greenland Ice Sheet Project 2 Depth-age Scale: Methods and Results. *Journal of Geophysical Research.* 102 (C12): 26411–26423.

10. Göktas, F. et al. 2002. A glacio-chemical characterization of the new EPICA deep-drilling site on Amundsenisen, Dronning Maud Land, Antarctica. *Annals of Glaciology.* 35 (1): 347–354.

11. Moore, J. C., H. Narita and N. Maeno. 1991. A Continuous 770-Year Record of Volcanic Activity from East Antarctica. *Journal of Geophysical Research.* 96 (D9): 17353–17359.

12. Meese, The Greenland Ice Sheet Project 2 Depth-age Scale: Methods and Results.

13. Seely, P. H. 2003. The GISP2 Ice Core: Ultimate Proof that Noah's Flood Was Not Global. *Perspectives on Science and Christian Faith.* 55 (4): 252–260.

14. Meese, The Greenland Ice Sheet Project 2 Depth-age Scale: Methods and Results.

15. Oard, M. J. 2005. *The Frozen Record.* Santee, CA: Institute for Creation Research, 8.

16. Hebert, J. 2014. Ice Cores, Seafloor Sediments, and the Age of the Earth: Part 2. *Acts & Facts.* 43 (7): 12–14.

17. The GISP2 timescale (as of May 13, 2014) has been archived on the National Oceanic and Atmospheric Administration National Climatic Data Center website at ncdc.noaa.gov.

18. Meese, The Greenland Ice Sheet Project 2 Depth-age Scale: Methods and Results, 26417-26419.

19. Ibid.

20. Ibid.

21. Southon, J. 2002. A First Step to Reconciling the GRIP and GISP2 Ice-Core Chronologies, 0–14,500 yr B. P. *Quaternary Research.* 57 (1): 32–37.

22. Meese, The Greenland Ice Sheet Project 2 Depth-age Scale: Methods and Results.

Chapter 5

1. Williams, L. D. 1979. An Energy Balance Model of Potential Glacierization of Northern Canada. *Arctic, Antarctic, and Alpine Research.* 11 (4): 443-456. Cited in Oard, M. J. 1990. *An Ice Age Caused by the Genesis Flood.* El Cajon, CA: Institute for Creation Research, 6–7.

2. Vardiman, L. 2008. Ice Age Glaciers at Yosemite National Park. *Acts & Facts.* 37 (3): 6.

3. Vardiman, L. and W. Brewer. 2010. Numerical Simulation of Precipitation in Yosemite National Park with a Warm Ocean: A Pineapple Express Case Study. *Answers Research Journal.* 3: 23–26.

4. Vardiman, L. and W. Brewer. 2010. Numerical Simulation of Precipitation in Yellowstone National Park with a Warm Ocean: Continuous Zonal Flow, Gulf of Alaska Warm, and Plunging Western Low Case Studies. *Answers Research Journal.* 3: 209–266.

5. Bradley, R. S. 1988. The Explosive Volcanic Eruption Signal in Northern Hemisphere Continental Temperature Records. *Climatic Change.* 12 (3): 221–243.

6. Morris, J. D. 2012. Volcanoes of the Past. *Acts & Facts.* 41 (6): 15.

7. Zielinski, G. A. et al. 1996. A 110,000-Yr Record of Explosive Volcanism from the GISP2 (Greenland) Ice Core. *Quaternary Research.* 45 (2): 109–118.